Josephine Mold

TESSELLATIONS

CAMBRIDGE : AT THE UNIVERSITY PRESS 1969

Published by the Syndics of the Cambridge University Press
Bentley House, 200 Euston Road, London N.W.1
American Branch: 32 East 57th Street, New York, N.Y. 10022

© Cambridge University Press 1969

Library of Congress Catalogue Card Number: 66–16668
Standard Book Number: 521 07420 7

Printed in Great Britain by Jarrold & Sons Limited, Norwich

The Romans decorated their buildings and towns with mosaic floors and pavements made of very small tiles called TESSELLAE. From this comes our word TESSELLATION, used to describe ways of filling space. The study of these is one of the bridges between mathematics and art. Such decorations have been used for centuries but today artists, architects and designers are making more and more use of simple geometric shapes and the ways in which they fit together.

Regular Polygons

Have you ever helped anyone to tile a wall or a floor? What shape were the tiles? Probably they were SQUARES—most tiles are. Why is that? Are there any square tiles where you are now, or outside?

Fig. 1

The photograph opposite shows a modern station booking-hall. You can see where there are square tiles. What shapes have been used to decorate the ceiling? Draw one of these shapes on thin card. Cut it out and draw round it to make a tessellation like that ceiling. (If you have any difficulty with this the activity on page 8 should help.)

A point where the corners meet is called a VERTEX. How many hexagons meet at each vertex? What does this tell you about the size of each corner of a regular hexagon?

What can you say about the size of a corner of a square by looking at a vertex of a tessellation of squares?

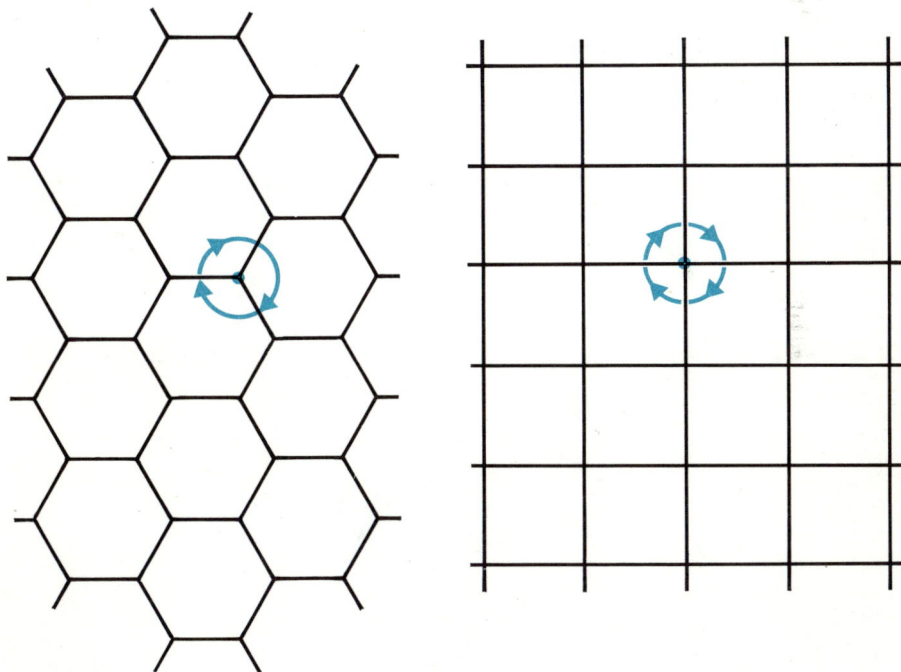

Fig. 2

Both the shapes used so far are called REGULAR POLYGONS. The word 'polygon' is used to describe any flat, straight-sided shape. Do you know any other words beginning with the prefix 'poly'? A large dictionary has hundreds of them. What does the prefix mean? These two particular shapes are also called 'regular' because each one has all its sides the same length and all its angles the same size.

Which of the shapes below are *regular* polygons?

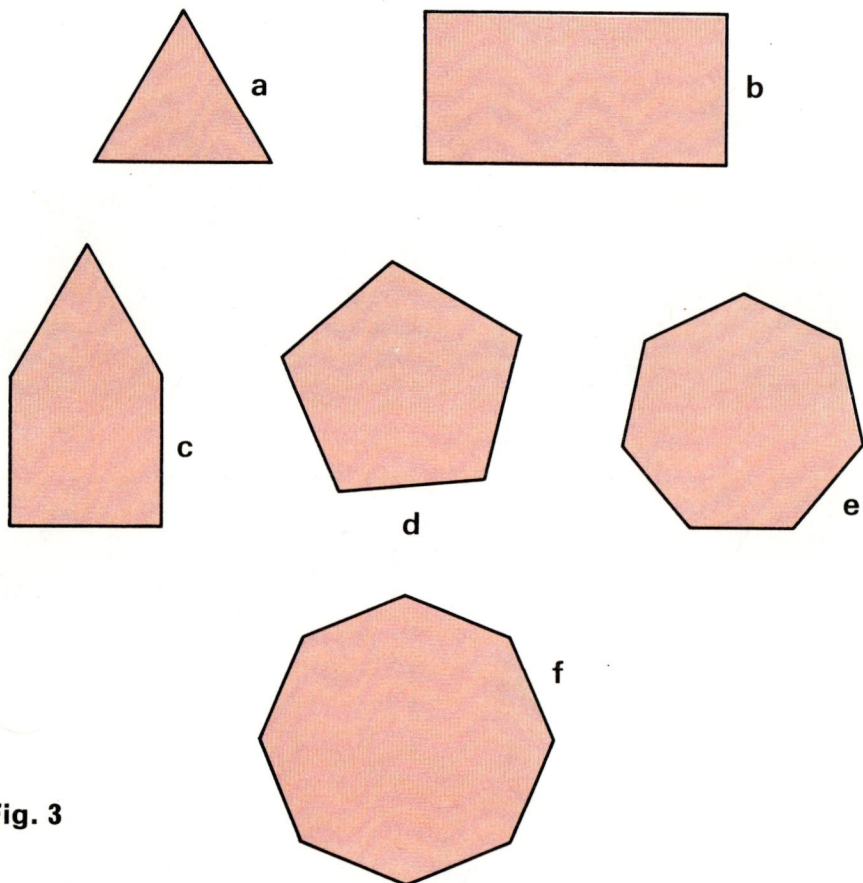

Fig. 3

Use tracing paper to find out which of the regular polygons in Fig. 3 can be used, without another shape, to make a tessellation.

Draw a set of diagrams to explain why only *one* of the *four* is suitable and explain why, out of all the possible regular polygons, only *three* (equilateral triangle, square and regular hexagon) will tessellate *on their own*.

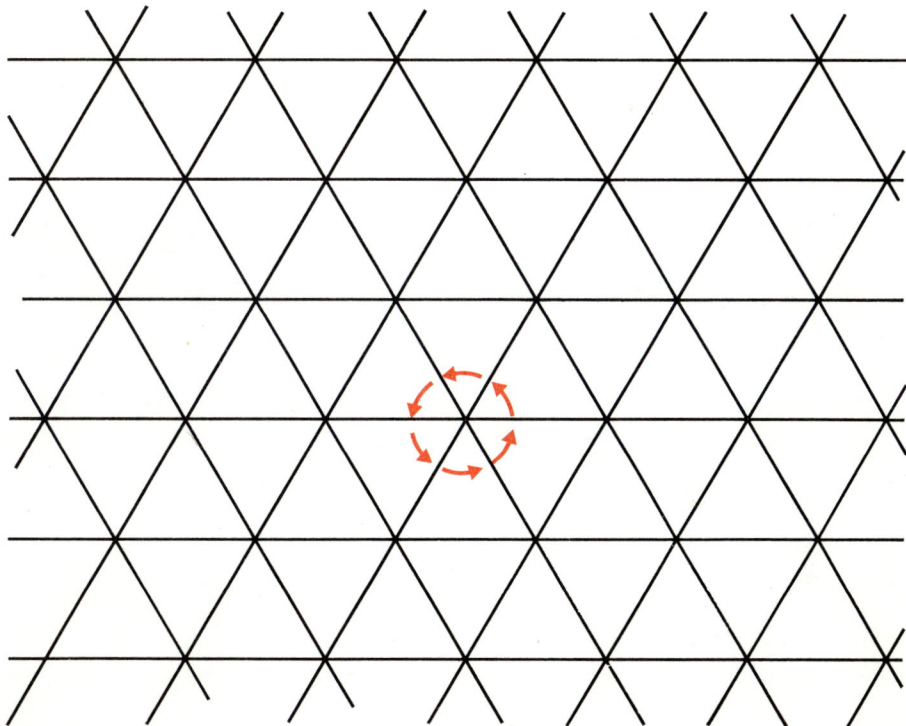

Fig. 4

What can you say about the size of each angle of an equilateral triangle by looking at each vertex in the tessellation? How does this compare with a regular hexagon? Can you see a tessellation of regular hexagons in Fig. 4?

Tessellations from Circles and Straight Lines

Take a large sheet of drawing paper and use compasses to make this circle pattern all over it.

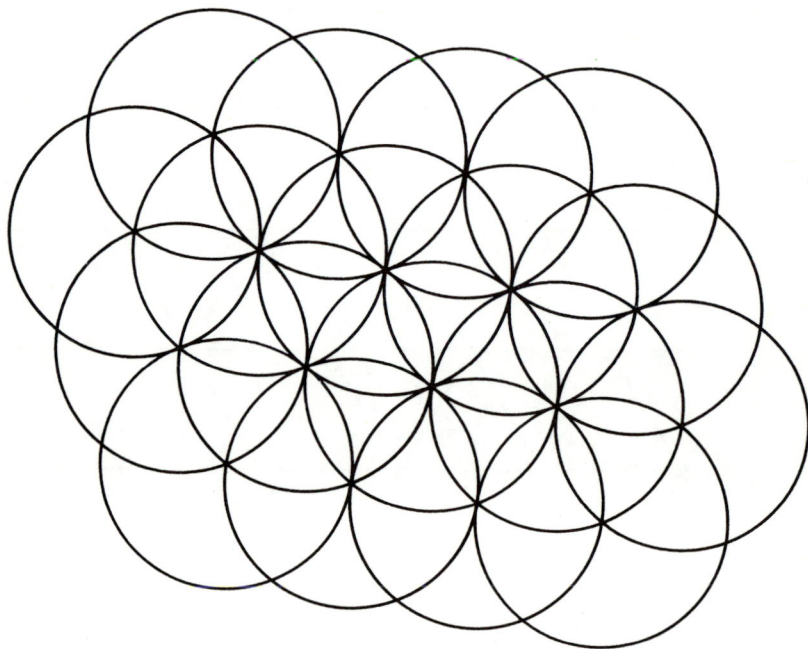

Fig. 5

Choose special points from the pattern to join up to make a tessellation of equilateral triangles.

Can you find other points which give a different size equilateral triangle? There are quite a number of different ways.

Now choose points which make a tessellation of regular hexagons.

Somewhere else on your pattern try to find points which give a tessellation of RECTANGLES. Can you find rectangles of a different size?

Now join points to make a tessellation of RIGHT-ANGLED TRIANGLES; then ISOSCELES TRIANGLES (triangles with two equal sides); then triangles with no right-angles and no equal sides.

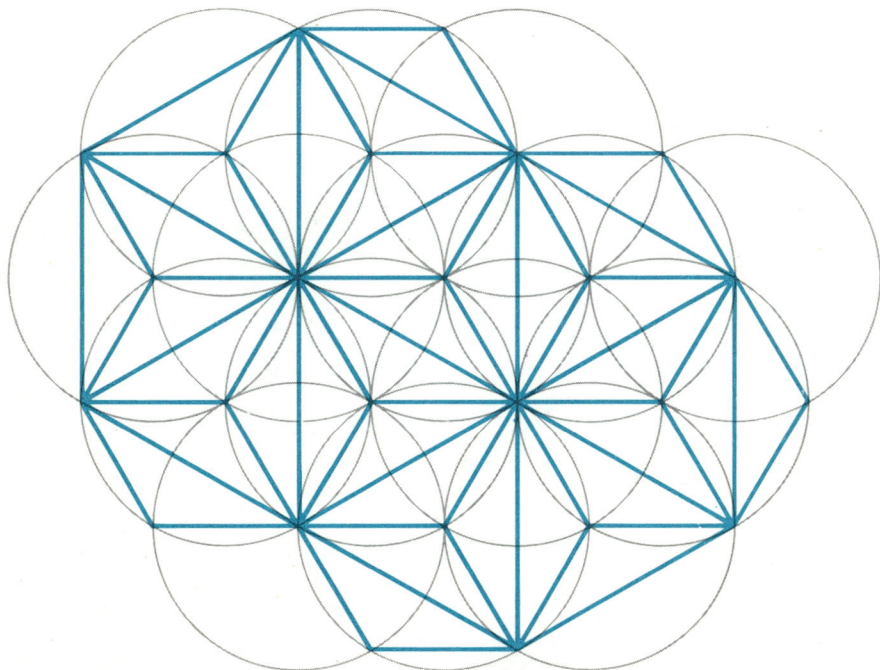

Fig. 6

Fig. 6 shows a tessellation of isosceles triangles. Did you get that one?

Find other tessellations by joining particular points on the circle pattern.

Use a ruler to draw a set of straight lines a ruler-width apart all over a sheet of drawing paper. Then twist the ruler round a bit and draw another set of parallel lines, the ruler-width apart, over the first.

What do the two sets of lines make? What would happen if you changed the angle between the two sets? What happens if the angle is a right angle?

Colour one of the shapes formed and draw its DIAGONALS. Draw the diagonals of all the shapes.

What do the diagonals form? Compare your results with someone else's. What would happen if your original two sets of lines were at right angles? Draw a diagram to show it.

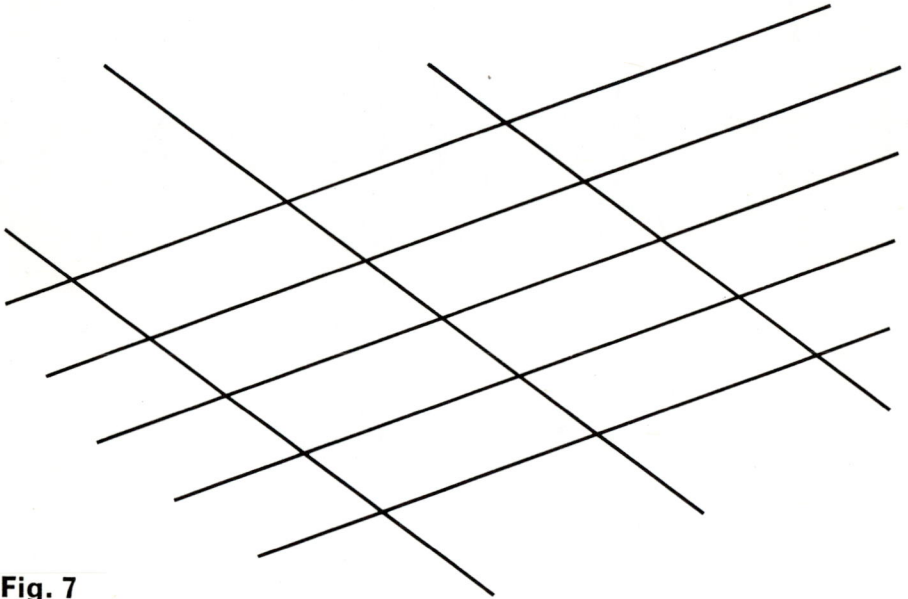

Fig. 7

Fig. 7 shows two sets of lines drawn with rulers of *different* widths. What difference has this made to the original shapes? Do this yourself and find out what happens if you put in the diagonals.

Redraw Fig. 7 making the two sets of parallel lines cross at right angles. What tessellations do the *diagonals* make?

By making the distance between the parallel lines change and by changing the angle and choosing various special points to join you can make a great variety of tessellations. Experiment for yourself and keep your results to use later. Fig. 8 may provide some ideas.

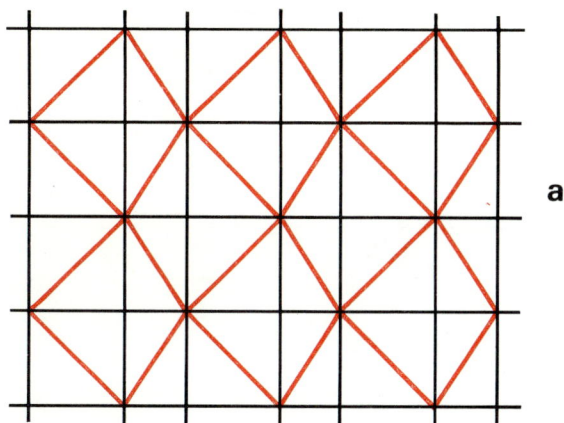

a

Fig. 8

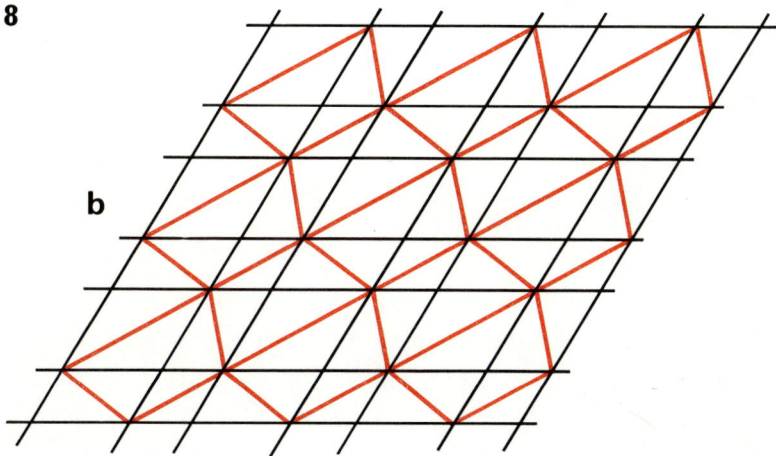

b

What shapes do you make if you only put in the longest diagonal of each parallelogram in Fig. 7? By varying the angle and the distance between the parallel lines you can make a tessellation of any triangle you like.

Try it by drawing any triangle, extending the sides and putting in the parallel lines. (See Fig. 9.)

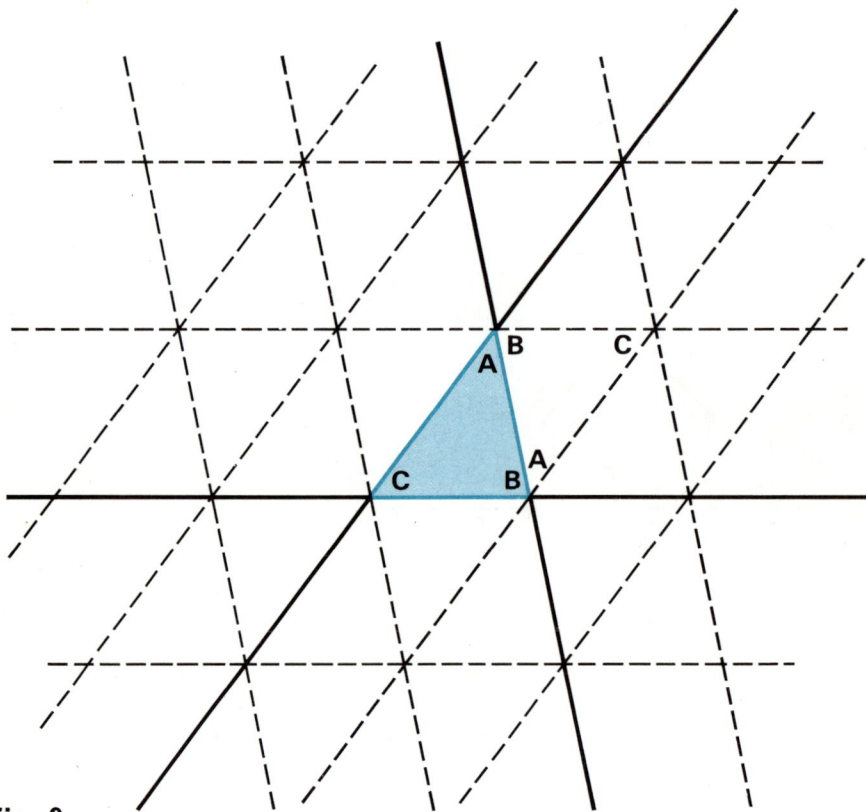

Fig. 9

Label the three angles of the triangle A, B and C and mark in where the same size angles appear at other places in the diagram. By looking at the angles all round one vertex, what can you say about A and B and C added together?

Fig. 8(b) on page 11 shows how QUADRILATERALS (four-sided polygons) can be obtained from two sets of parallel lines. Notice the pattern of distances between the lines. Draw any quadrilateral and mark in the diagonals. Now build up the two sets of parallel lines and complete the tessellation.

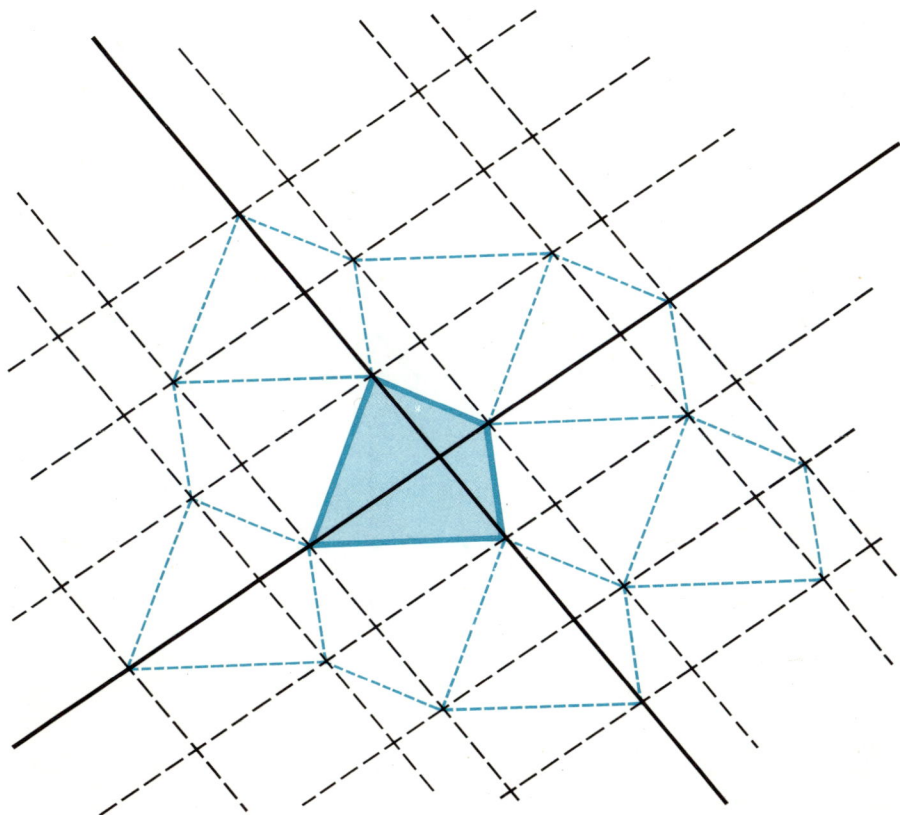

Fig. 10

Colour the quadrilaterals in two colours to show up clearly how they are arranged in the tessellation. Work out the sum of the angles at the corners of a quadrilateral in the same way as for the triangles on the opposite page.

Combining Shapes

One of the regular polygons on page 6 was an OCTAGON (eight sides). It would not tessellate on its own but you probably noticed that the spaces left were squares. So if you use octagons *and* squares you can make a new tessellation.

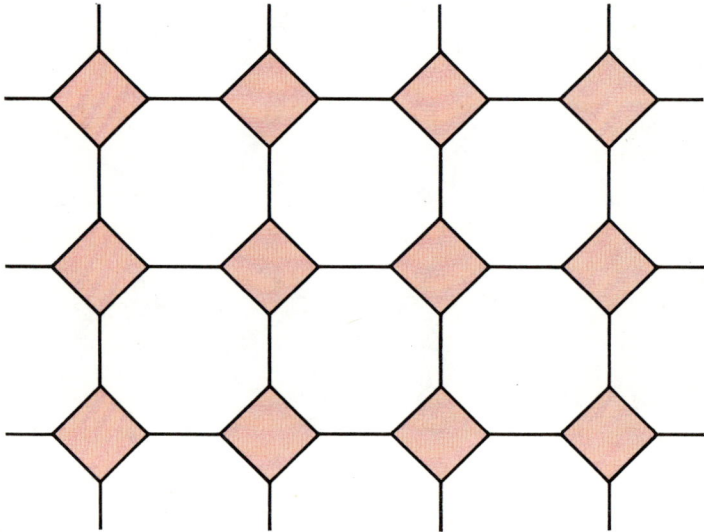

Fig. 11

Make a larger version of this to use later.

This particular pattern was used quite frequently in Victorian England and may still be seen in decorative tiled pathways and passages, windows and wallpapers of some old houses. Look out for it.

The three regular polygons which tessellate on their own can also be used in combination. Make a number of paper equilateral triangles, squares and regular hexagons all with the same length side.

Experiment with your cut-out polygons to find tessellations of two or more of the shapes. Remember it must be possible to see how each could be continued on and on . . .

You should be able to find five different examples. Make a good drawing of each one.

Earlier in this book you found that each angle of a square must be a right angle because four, equal-sized angles fitted the space around each vertex of the tessellation. In the same way you saw that the angle of a regular hexagon was one-third of a turn. One whole turn is called 360 degrees (°). How many degrees would each angle of a regular hexagon be? Which regular polygon would have all its angles 60°? Use Fig. 11 to work out the angle of the corner of a regular octagon.

Fig. 12 shows a tessellation which includes a regular DODECAGON (twelve-sided polygon). Work out the angle at the corner of a regular dodecagon.

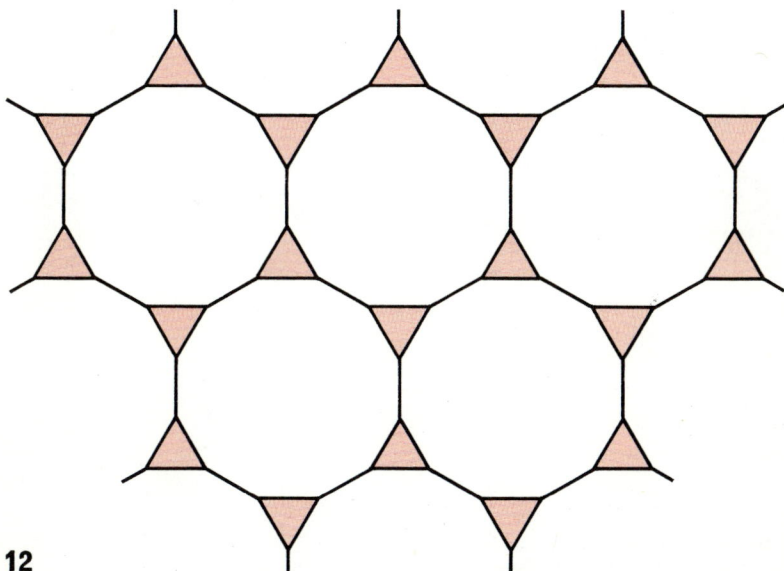

Fig. 12

Copy and complete the table below.

Draw a graph to illustrate the relation between the numbers in columns (1) and (2) and another for columns (1) and (3).

Name of polygon	(1) Number of sides	(2) Size of angle	(3) Total of angles
Equilateral triangle	3	60°	180°
Square	4	90°	360°
Regular hexagon	6		
Regular octagon	8		
Regular dodecagon	12		

Can you say anything about the angles of regular five, seven, nine, ten and eleven-sided polygons from your graphs?

Do you think that you should join up the points on your graphs? Give reasons for your answers and discuss them with your teacher.

You may have begun to notice some of these tessellations in buildings, on material, wallpaper, floor-covering or wrapping-paper. A tessellation of hexagons appears in honeycomb and chicken wire, sometimes in a patchwork quilt. Hexagonal tiles are sometimes seen on walls and floors, and hexagonal paving stones in shopping centres and under bridges on some motorways. Look out for them.

The tessellation shown in Fig. 13 appeared, made of small clay tiles, on the floor of an old shop in Windsor. The builder had done a lot of decorative work at Windsor Castle in the reign of Queen Victoria and this probably accounted for the unusual tiles he had.

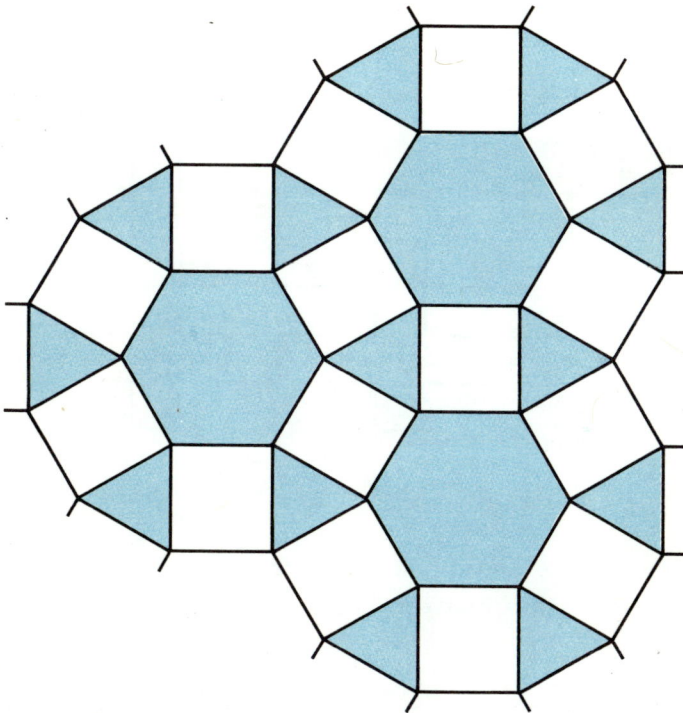

Fig. 13

Make your own collection of sample materials, floor-covering, wrapping-paper, wallpaper etc. Make sketches of tessellations you see which are too large to bring into the classroom.

Shapes from Squares and Triangles

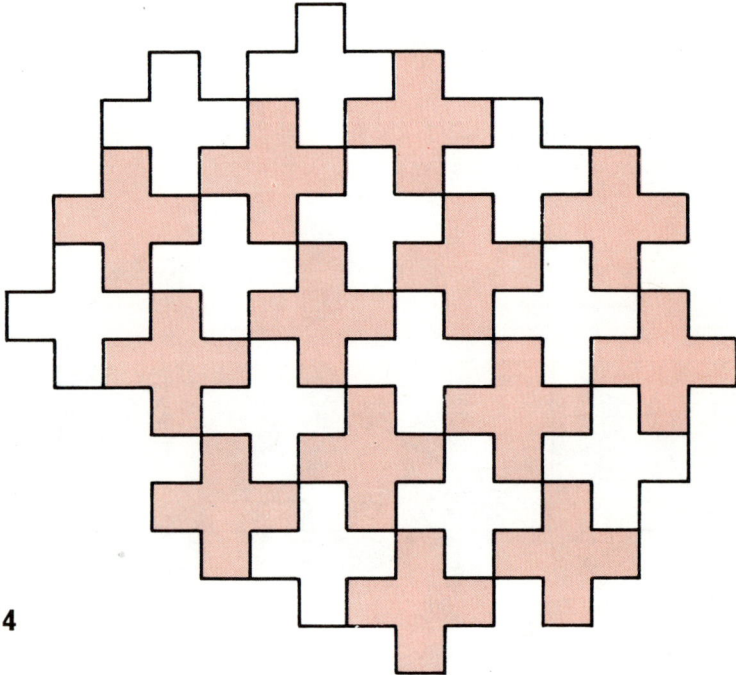

Fig. 14

Fig. 14 shows a tessellation of a cross shape made from five squares. There are many other five-square shapes such as . . .

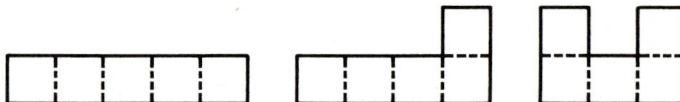

Fig. 15

These shapes are called PENTOMINOES. Notice that each square is joined to at least one other by a whole edge—there is no joining just by a corner. Find as many other pentominoes as you can and see which ones will tessellate. Can you find any which can be tessellated in more than one way? You can find out more about these shapes if you use the book *Cubes* in this same series. You can read about them also in the books numbered (7) and (13) in the book-list on page 32.

Whenever you do a jig-saw puzzle you are making a tessellation although the shapes are not all the same. Often they look rather like the shape of nine squares . . .

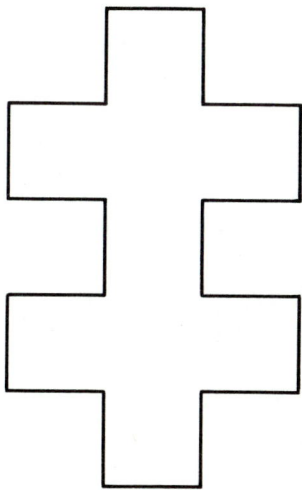

Fig. 16

Use this shape to make two different tessellations and colour each one in two colours to show up the difference in their construction.

It is also possible to build up shapes made of triangles in the same way and some of these will tessellate as well.

Some tessellations of triangular-based shapes are drawn below. Do any of them look three-dimensional to you? If so, can you explain why? Which ones are the same tessellation differently coloured? Notice what a big difference the colouring makes to what you see. Make some tessellations of shapes made from triangles for yourself.

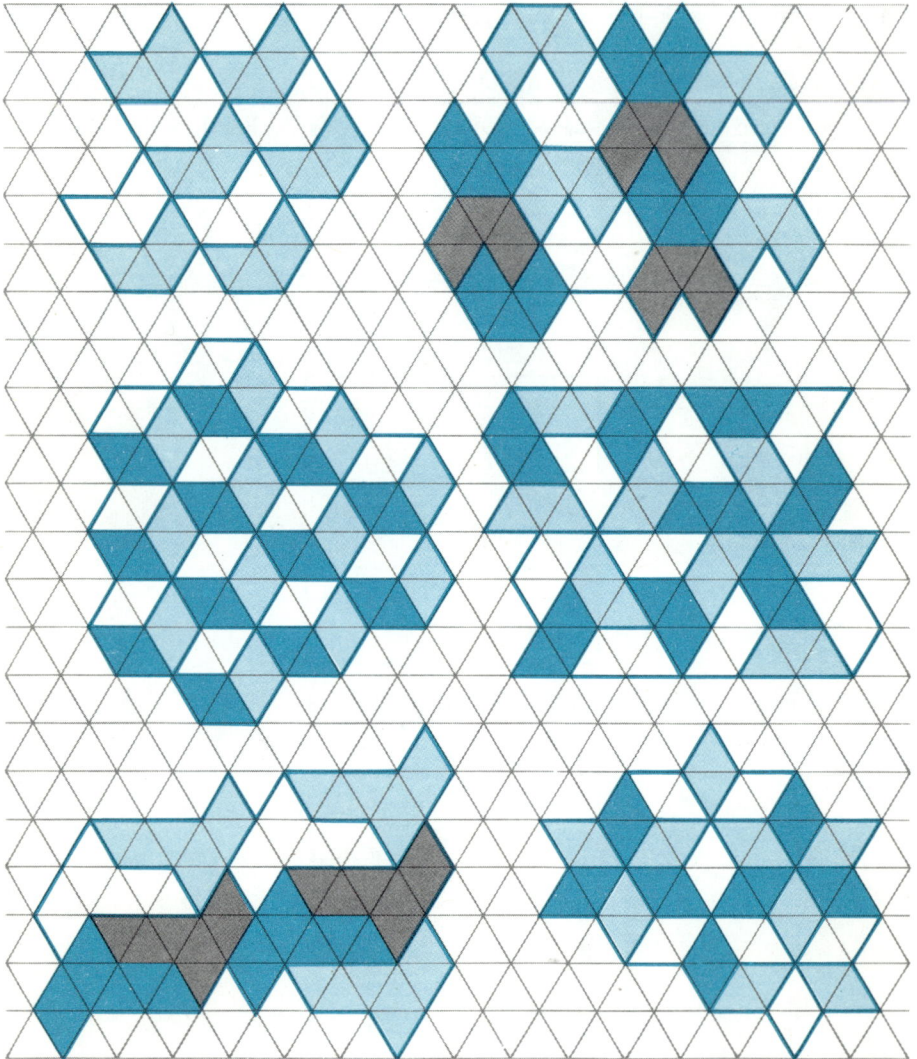

Fig. 17

Cut out two squares of different sizes and use them to build up a tessellation of which all the vertices are identical, having the same arrangement of squares around them.

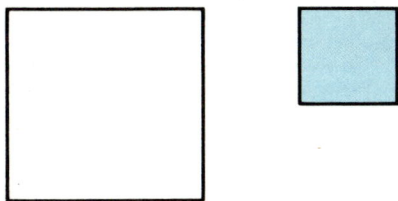

Fig. 18

Mark the centres of the larger squares. What tessellation is made by the centres? Can you discover any relationship between the area of the two squares and the area of the squares in the new tessellation? Discuss this with someone else.

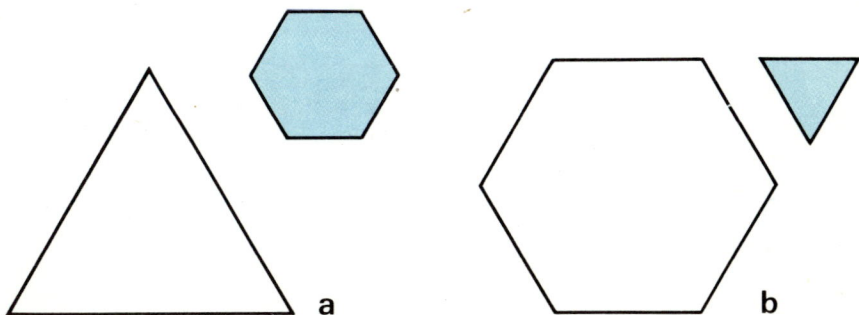

Fig. 19

Make tessellations from a large regular hexagon and a small equilateral triangle and then from a large equilateral triangle and a small regular hexagon—again with identical vertices in each case. Mark the centres of the larger shapes and look for area relationships. Do these relationships still hold if the hexagon and the triangle have the same length edge?

Symmetry

The two tessellations which can be made with the jig-saw piece which you used on page 19 are shown in Figs. 20 and 21.

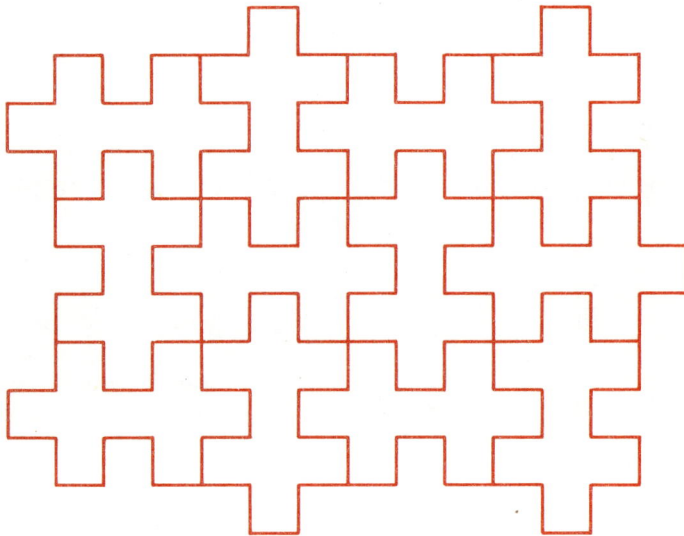

Fig. 20

How would you describe the differences between them?

Find the drawings you did and, taking each in turn, try to stand a small mirror somewhere on them so that as you look into the mirror the pattern seems to continue, undistorted, through it. Mark in all the possible LINES OF REFLECTION or AXES OF SYMMETRY, as they are called. Does a single shape from the pattern have any axes of symmetry? If so, how do they compare with those of the whole tessellation?

Now trace a small section of each tessellation. See if you can then fix one point on the tracing and gradually turn the paper round it until the tracing fits back on to the tessellation. How far did you have to turn the tracing round? Does it depend on the turning point? Mark in turning points which only need a 90° turn before the tracing fits again. Now mark, in another way, points which need a 180° turn. Which points have to have a whole turn? What shapes do your marked points make? Again compare your results with those for a single piece of the tessellation.

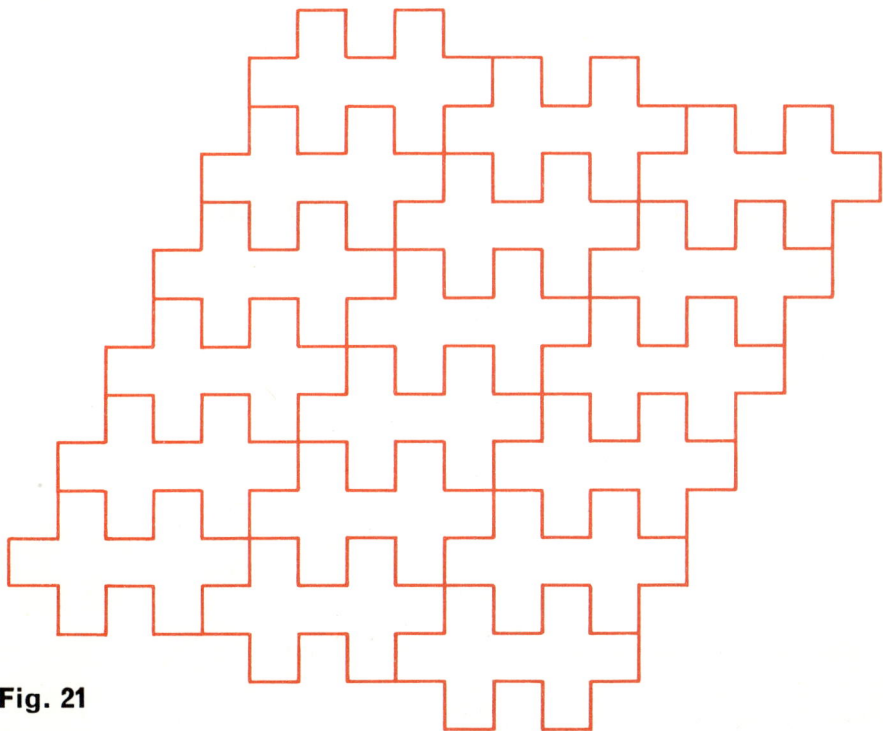

Fig. 21

While you were doing this you may have noticed that the tracing could be made to fit again without any turning at all—just by a straight slide in various directions. Investigate the possibilities.

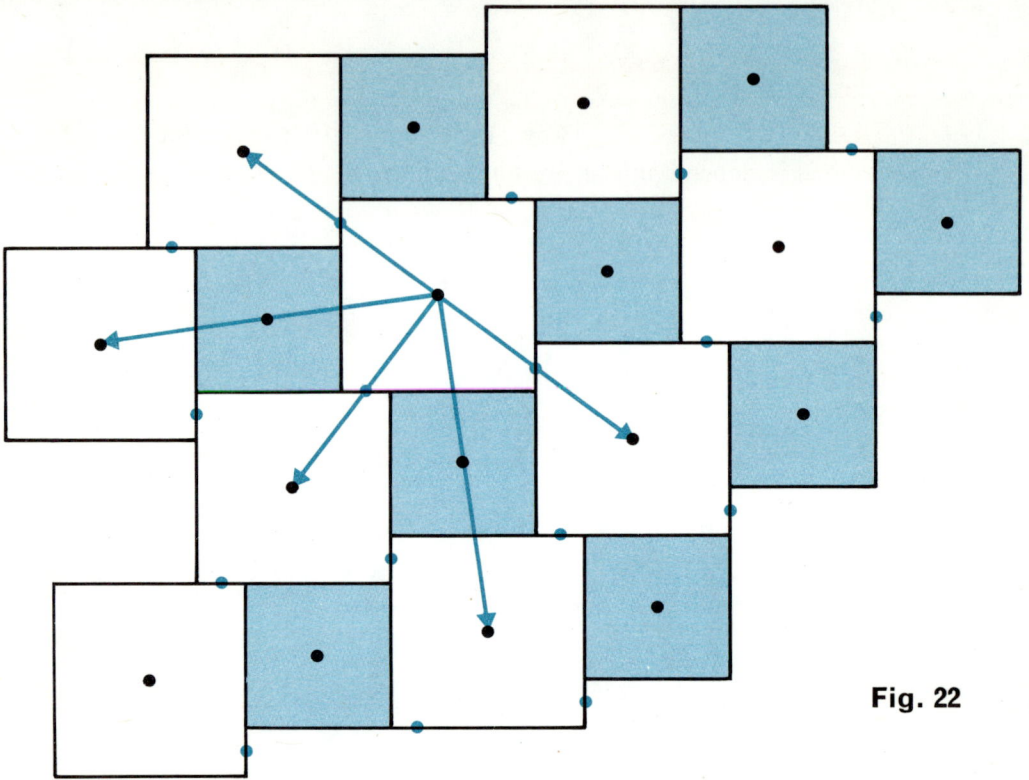

Fig. 22

Fig. 22 shows the tessellation you made from two different squares (page 21). The black dots are the CENTRES OF ROTATION where turns of 90° are enough to bring the tracing back on to the tessellation; the coloured dots are centres where 180° are needed. Notice the arrangement of the dots.

This particular tessellation has no lines of reflection.

The coloured arrows show *some* of the possible distances and directions of TRANSLATIONS (slides) which make tessellation and tracing coincide.

Investigate two or three of your other tessellations in the same way. These reflections, rotations and translations are called the SYMMETRY OPER-ATIONS of the tessellations.

Duality and Colouring

If you have already used the book called *Solid Models* in this series then you will have met the idea of duality.

Take any one of your tessellations of regular polygons and mark the centre point of each shape. Join the centre of each shape to the centres of the shapes which are edge to edge with it.

Have you made another tessellation? Is it a tessellation of a single shape or more than one shape? The second tessellation is called the DUAL of the first.

Fig. 23 shows one of the tessellations (in black) and its dual (in colour).

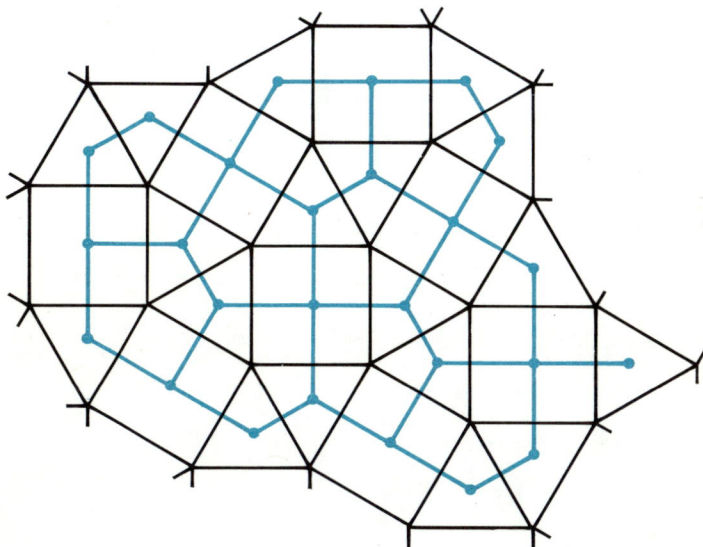

Fig. 23

The dual shown in Fig. 23 on the previous page is a tessellation of irregular, but identical, PENTAGONS (five-sided polygons). Which other duals were pentagons? How can you tell, from the original tessellation, whether or not a pentagon will appear and whether or not the shapes in the dual will be identical? Discuss the answers with your teacher.

Compare the symmetry operations of the duals with those of the original tessellations.

You have already coloured some of your tessellations to show up their construction. There are several different ways of doing it. You might choose to colour each type of shape differently so that they can be picked out easily. In a tessellation of one type of shape you might colour shapes according to which way round they are (as you did the quadrilaterals on page 13, for example). A third possibility is to make sure that no two shapes which share an edge have the same colour. This last way is how map-makers colour counties or states, for example.

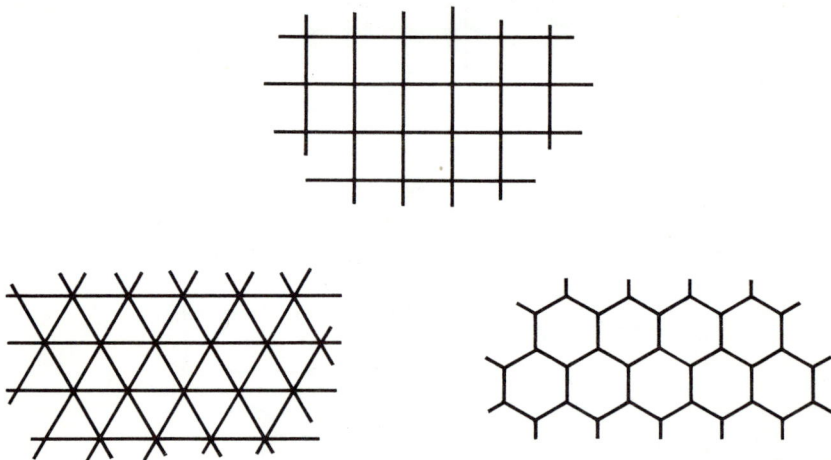

Fig. 24

What is the *least* number of colours, under the 'map-maker's rule', needed to colour each of the three regular tessellations which are drawn in Fig. 24? Try it on your own copy. Colour more of your tessellations with the *least* possible number of colours under this rule.

You will have found that you only need two, three or four colours each time. Fig. 25 shows one of each type. How can you tell straightaway from any tessellation if two colours will be enough?

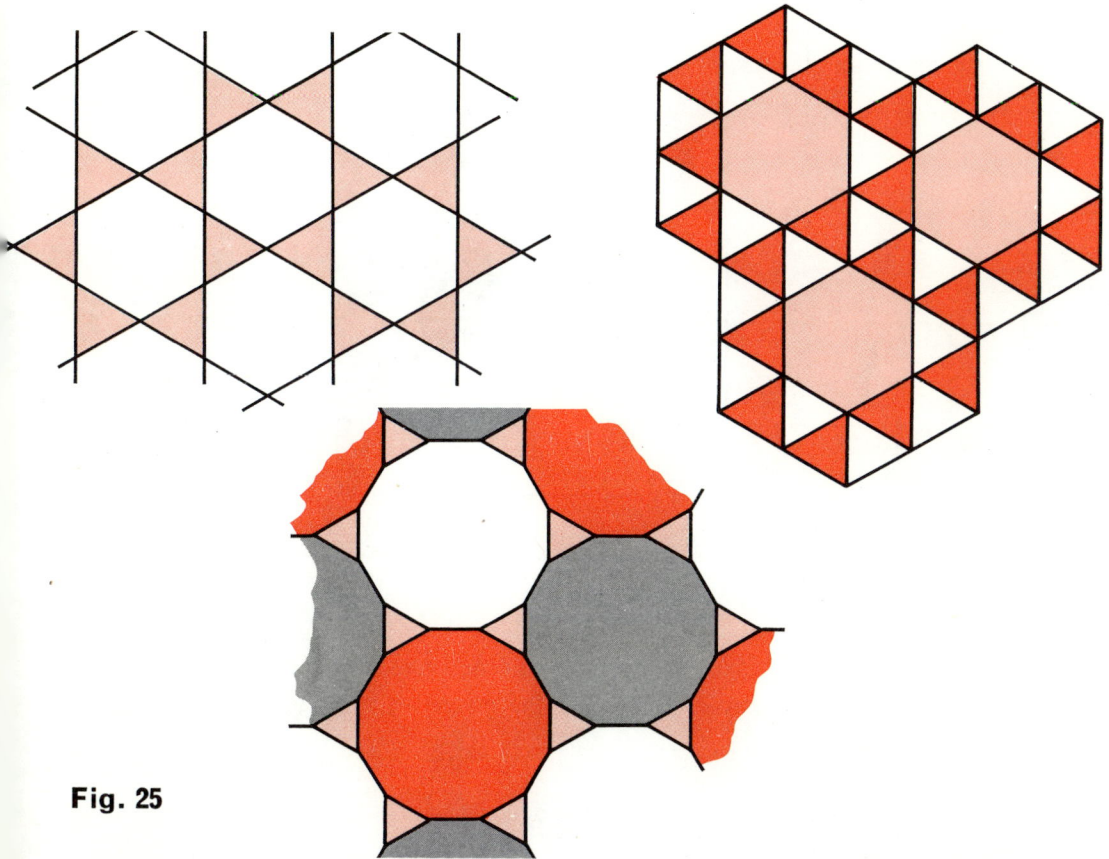

Fig. 25

There are many colouring problems, not all yet solved, in which you might be interested. You can find out more about them if you read books (2) and (10) in the list on page 32.

Curved Tiles

Did you notice the cover-design of this book? The pattern is an ancient Moorish one which is still often used today. It is possible, but expensive, to buy tiles this shape. There is a floor-covering of this design which is a good imitation of actual tiles and you can buy square floor tiles, like Fig. 26, which make the pattern when fitted together.

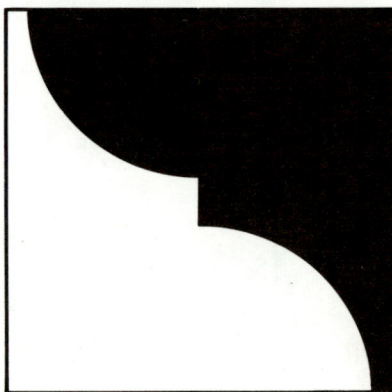

Fig. 26

Try to design some square tiles which give a tessellation of a single shape with curved edges when fitted together.

It is not difficult to extend the idea to other shapes. The equilateral triangle in Fig. 27, for example, will make the curved-shape tessellation shown in Fig. 28, opposite.

Fig. 27

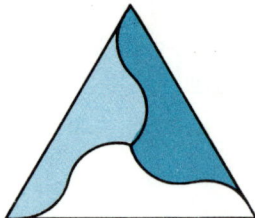

28

Fig. 28

If you really had to tile a floor to get a curved effect, then this method would probably be best, as the square and triangular shapes pack and transport easily. (Why?) If you plan to design material, wallpaper or wrapping-paper, however, you can easily adapt the shapes in the tessellations themselves. The basic quadrilateral tessellation could be turned into two shoals of fish swimming in opposite directions!

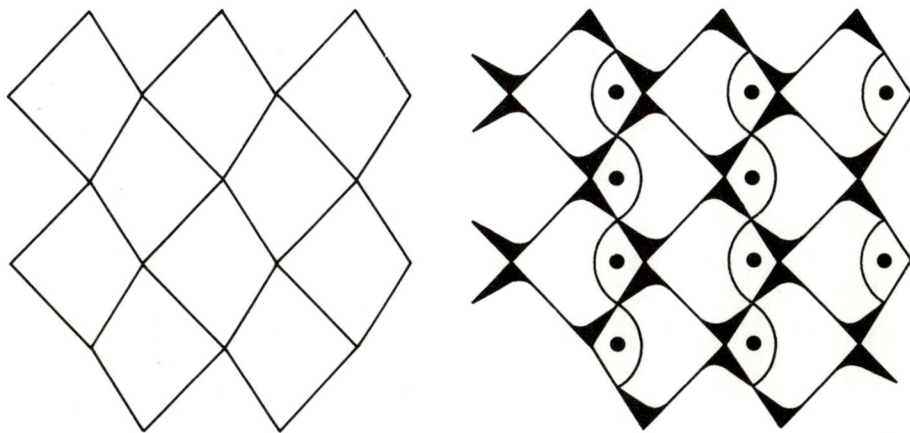

Fig. 29

Turn your quadrilateral tessellation into something curved. Remember that whatever you take from one shape must be attached to another.

An equilateral triangle can have a simple arc cut out from and added to each side, or it may be made more complicated. Fig. 30 gives two examples; in both these motifs an 'S' shape has been drawn on each side, crossing at the centre.

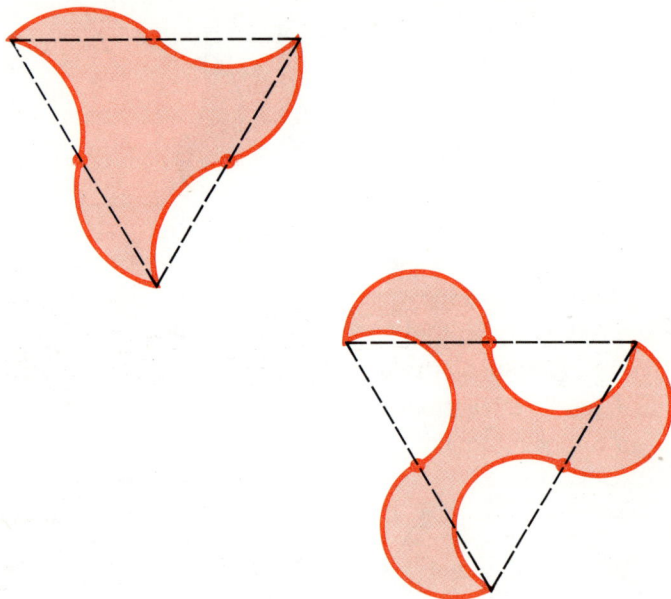

Fig. 30

Enlarge and cut these motifs from card and draw round them to make tessellations. Invent others based on some of your original tessellations.

There are two photographs (Fig. 31), on the page opposite, of drawings by the artist Maurits Escher. They are taken from book (1) in the list on page 32. Maybe there is a copy of this book in your school or local library; try to see one, as there are a lot of clever drawings in it. These two are from a whole series based on various tessellations. It is easy to see how the first is constructed. The second is not so simple: like the first it uses equilateral triangles but it also uses the hexagons found in it. Work out how it is done and create some others for yourself.

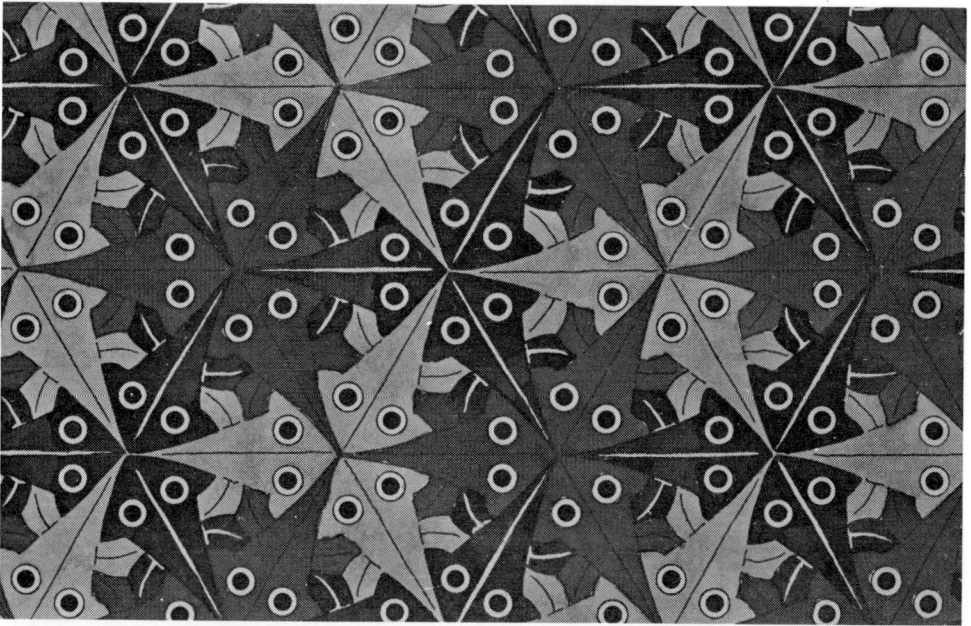

Fig. 31

first introduction to a topic which, apart from its own great intrinsic interest, has connections with art, science and more mathematics. The following book-list gives some of the books which would be useful for classroom reference by teachers and pupils who wish to pursue some of the questions raised. Those marked 'T' are mainly for teachers.

(1) *The Graphic Work of M. C. Escher* (Oldbourne Press; Duell, Sloane and Pearce)
(2) *Mathematical Recreations and Essays* by W. Rouse-Ball (Macmillan)
(3) *Pattern and Shape; The Development of Shape; Shapes we need*: all by Kurt Rowlands (Ginn)
(4) *Pattern Design* by Lewis S. Day (Batsford)
(5) *Mathematical Models* by H. M. Cundy and A. P. Rollett (Oxford University Press)
(6) *The Third Dimension in Chemistry* by A. F. Wells (Oxford University Press)
(7) *Mathematical Puzzles and Diversions* by M. Gardner (Scientific American series, Simon and Schuster; Bell; Pelican)
(8) *Mosaics* by D. Stover (Houghton Mifflin; Heinemann Educational)
T (9) *Groups and their Graphs* by I. Grossman and W. Magnus (Random House)
(10) *Mathematics* (*Life* Science Library)
T(11) *Regular Figures* by L. Fejes Toth (Macmillan)
T(12) *Symmetry* by H. Weyl (Princeton University Press; Oxford University Press)
(13) *Polyominoes* by S. W. Golomb (Scribners; Allen and Unwin)
T(14) *Symmetry Groups* (Association of Teachers of Mathematics Pamphlet)

ACKNOWLEDGEMENTS

We should like to thank the following for allowing us to reproduce illustrations: Bakelite Xylonite Ltd for the photograph of Chichester station (Fig. 1) and Erven J. J. Tijl N.V. for the drawings by Maurits Escher (Fig. 31).